The Diffusion of Crude Petroleum through Fuller's Earth.

DISSERTATION.

SUBMITTED TO THE BOARD OF UNIVERSITY STUDIES OF
THE JOHNS HOPKINS UNIVERSITY IN CONFORMITY
WITH THE REQUIREMENTS FOR THE DEGREE OF
DOCTOR OF PHILOSOPHY.

By

OSCAR ELLIS BRANSKY.

EASTON, PA.:
ESCHENBACH PRINTING COMPANY
1911.

ACKNOWLEDGMENT.

To President Remsen, and Professors Morse, Jones, Renouf, and Acree, the author is greatly indebted for valuable instruction in the lecture room and laboratory. The author wishes, in particular, to express his gratitude to Dr. Gilpin, under whose guidance this investigation has been pursued, and to Dr. Day, of the United States Geological Survey. Thanks are also due to Prof. Swartz for important suggestions.

CONTENTS.

Acknowledgment... 3
Introduction... 5
Object of this Investigation............................. 14
Experimental.. 14
    I. Relative Amounts of Oil Lost in Heated and Unheated
        Fuller's Earth.................................... 14
   II. The Diffusion of Benzene in Solution through Fuller's
        Earth.. 18
  III. The Fractionation of Crude Petroleum.................. 31
  IV. Chemical Examination of the Fractionated Oils........... 49
   V. Selective Action of Fuller's Earth...................... 53
Summary... 56
Biographical.. 58

The Diffusion of Crude Petroleum through Fuller's Earth.

INTRODUCTION.[1]

It is a well-established fact that the petroleum obtained from the sandstones of the Upper Devonian and Mississippian periods, generally known as the Pennsylvania oil, differs markedly from the natural oil found in the Trenton limestone, usually designated as the Ohio oil, or Trenton limestone oil. Both of these oils, in turn, are distinctly different from the petroleum occurring in the loose sands and soft shales of California. The unconsolidated tertiary clays, sands, and gravels in the southern United States, particularly in Texas, yield another variety of petroleum, characterized by properties more or less different from any of the preceding oils.

Not only do these differences exist between oils found in separate regions, but extreme variations in color and specific gravity, as well as in chemical composition, often occur between those of neighboring localities. On the other hand, close resemblances abundantly occur between petroleums of sections widely removed from each other. Some of the South American and many of the European oils, for instance, have been found to possess properties very similar to those of the oils of the southern United States; while the oil from the Corniferous limestone of Canada closely resembles the Ohio petroleum.

These variations in the oils of the United States and other countries have been carefully studied by many investigators. Such noted workers as Warren, Storer, Mabery, Pelouze, Cahours, Schorlemmer, Beilstein, Markownikoff, Engler, and Kurbatoff have devoted their lives to this subject. The question that naturally arises in connection with these variations is: Are these differences fundamental? Is the Pennsylvania petroleum as distinctly different from the Ohio oil as one chemical compound is from another? In answer to these

[1] This research was aided by a grant received from the C. M. Warren Committee of the American Academy of Arts and Sciences.

questions, the following extract from a paper read by Mabery[1] in 1903 before the American Philosophical Society is of considerable importance: "Now, after years of arduous labor, I have reached the conclusion that petroleum from whatever source is one and the same substance, capable of a simple definition—a mixture of variable proportions of a few series of hydrocarbons, the product of any particular field differing from that of any other only in the proportion of the series and the members of the series." The evidence supporting this declaration has been and is accumulating constantly, and, at the present time, this view is generally accepted.

If petroleum, then, is one and the same substance, how can the extreme variations between the American oils be explained? Were the causes operating in the formation of the Pennsylvania oil, almost barren of sulphur and nitrogenous bodies, different from those acting in the production of the sulphur-bearing oils of Ohio, or the heavy sulphur and nitrogenous oils of California?

To account for the formation of crude petroleum, two views, as is well known, the organic and inorganic, have been advanced. The Pennsylvania oil, according to these theories, may have been formed either from organic or inorganic substances, or from both. It is as yet impossible, however, to state conclusively from which of these sources the oil was derived. It is apparent, therefore, that the differences between the Pennsylvania and the Ohio, Texas, and California oils cannot be explained upon the assumption that the former was formed from organic remains, while the latter were produced from inorganic matter, or *vice versa*. If, however, crude petroleum is organic in origin, it may have been formed either from vegetable or from animal remains. The following discussion is based upon the assumption that the above-mentioned oils were derived from an organic source.

It has been suggested that the differences between these oils may be accounted for by assigning a vegetable origin to the Pennsylvania oil, and an animal origin to the others.

[1] P. Am. Phil. Soc., 1903.

7

Mabery[1] states that "It would seem that the small proportion of these bodies[2] in the Pennsylvania oil, as compared with the larger proportions in the limestone oils and California oils should be strong evidence in favor of a different origin, that the Pennsylvania oil came from organic vegetable remains, which should permit of the small amounts of sulphur and nitrogen compounds from this class of oils." Newberry, Peckham, Orton, and other geologists also favor the view that the Pennsylvania oil is of vegetable origin, and is derived from the organic matter of the bituminous shales of the Devonian period.

The facts which have led to the association of this oil with a vegetable source are, first, that the oil is of a different character from the limestone oils of Ohio, and those of Texas and California; second, that the Pennsylvania petroleum is found in strata that bear but few fossils; third, the belief that the Chemung and immediately overlying formations are barren of animal organic remains; fourth, the existence of large quantities of microscopic fossils, whose origin many believe to be vegetable, in the black shales of the Lower and Middle Devonian periods, to which formations many investigators are inclined to refer the origin of the Pennsylvania oil.

It is generally recognized that the Pennsylvania oil differs markedly from the Ohio, Texas, and California oils. Investigation has shown that the former contains a much larger proportion of the paraffin hydrocarbons, and a much smaller percentage of benzene and unsaturated hydrocarbons and sulphur and nitrogenous bodies, than the latter oils. It is further generally admitted that the Pennsylvania oil was not formed *in situ*. These two facts have aided strongly in assigning a vegetable origin to this oil. To what strata, then, should the source of the oil be referred? The great coal formations of Pennsylvania, lying above the Chemung, seem, at a first glance, to offer a solution of this problem. It is a notable fact, however, that these formations have not, up to the present time, been connected, either chemically or geologically, with the Pennsylvania oil. The possibility exists, however,

[1] P. Am. Phil. Soc., 1903.
[2] Reference is made to the sulphur, nitrogen, and oxygen compounds in petroleum.

that it was formed from vegetable remains in the Carbonifer-
ous formations above, and that then, by downward diffusion,
it reached its present position in the Chemung. This view
rests upon the physical fact that a liquid diffuses by the force
of capillarity in all directions, downward as well as upward.
Little attention has been given to this possibility, but it seems
to deserve a very careful study. Owing, however, to the uni-
versal association of water under hydrostatic pressure with
natural oil and gas, the migration of the latter is generally
upward. This fact is attested by the accumulation of oil
in anticlinal folds when water is present, and by the existence
of the remarkable gushing oil wells. That the Pennsylvania
oil, if not formed *in situ*, ascended to its present location seems,
therefore, more probable.

In what strata below the Chemung, then, was the oil originally produced? It has been previously mentioned that a
number of investigators refer the source of the oil to the black
shales of the Lower and Middle Devonian periods. The organic matter of these shales is composed largely of microscopic sporangites, which suggest the existence, according to
Orton, of masses of floating vegetation, or Sargasso seas.
According to this view, the origin of the Pennsylvania oil is
vegetable in character, and its primitive abode was in the
shales of the Devonian age lying below the Chemung sandstone, to which it ascended under the influence of natural
agencies. Another origin, animal in character, may be as-
signed to this oil. This view is that the oil was formed in
the fossil-bearing strata of the Chemung age, and that it
diffused to the sandstone reservoirs in which it is now found,
and that during such a diffusion its original character was
changed. Prof. C. K. Swartz, of the Johns Hopkins University, who has made a critical study of the Chemung strata
in Maryland, informs us that fossil remains exist in considerable abundance in the strata of this age in Maryland and ad-
joining areas. In Pennsylvania, the corresponding strata
have been found to bear many fossils. It is possible that the
oil formed in these strata, and then diffused to the strata in
which it now exists and which are barren of fossil remains.

The evidence accumulated in this investigation seems to show that it is not necessary to assign a vegetable origin to the Pennsylvania oil to explain the differences between it and the oils of Ohio and California. It is clear from the results of this and other investigations that, when such oils as those of Ohio and California and Texas, which seem to be of animal origin, are allowed to diffuse through such porous media as fuller's earth, they yield oils very similar to those of Pennsylvania. By assuming, therefore, that the Pennsylvania oil migrated from some primitive source, in which it may have been formed from animal remains, through shales, limestones, and sandstones, its peculiar character can be understood.

Whatever the original home of the oil, it seems probable that it migrated to its present location from some place below. It is with the changes occurring in crude petroleum as a result of such a migration through porous strata that the present investigation is primarily concerned.

In 1897, Dr. David T. Day,[1] from his own observations, and those of Dr. John N. MacGonigle, proposed the view that the Pennsylvania oil, at some past time, possessed properties very similar to those of the Ohio oil, but that in its migration to its present abode from regions below, its character was changed to its present condition.

Guided by this view, he conducted, in the laboratories of the United States Geological Survey, an investigation into the changes occurring in crude petroleum when allowed to diffuse through porous media, such as fuller's earth. He demonstrated clearly that an oil resembling the light Pennsylvania oil could be readily produced in the laboratory from the heavier crude Ohio oil. Glass tubes were packed firmly with the dry earth, through which the crude oil diffused by its own force of capillarity. From the earth of the upper sections of the tubes, very light, in some cases colorless, oils were liberated by treatment with water; from the earth of the lower sections of the tubes, much darker and heavier oils were obtained.

[1] P. Am. Phil. Soc., 1897.

It will be observed that the fractionation is effected entirely by capillarity; oils with different surface tensions rise with different velocities through the capillary openings, such as the fine interstices and minute pores of the fuller's earth. A separation of the various constituents making up the complex of any one oil is thus produced. The view once held that this phenomenon is chemical was clearly disproved by Engler and Albrecht[1] in 1901, and later by other investigators.

Any medium, therefore, sufficiently fine-grained and porous to afford capillary spaces causes a separation of the constituents of any mixture, provided they possess different surface tensions. The compact sandstones, shales, and limestones that recur in many cycles throughout the earth's crust present an excellent medium for the separation of the constituents of such a complex mixture as petroleum. The force of capillarity, assisted by the hydrostatic pressure of the water occurring in the interior of the earth, acting over vast periods of time, is, it seems safe to state, sufficiently powerful to transport the oil from the lower regions to those above. That the condition, therefore, to cause such a migration, with the consequent fractionation of the original oil, are abundantly present, appears extremely probable.

Let us examine, now, the conduct of the constituents of petroleum subjected to such a fractionation. The members composing the natural oil may be grouped under the following general heads: paraffin, aromatic, unsaturated hydrocarbons, sulphur, nitrogen, and oxygen compounds. The behavior of the paraffin and unsaturated hydrocarbons will be considered first.

Dr. David T. Day early observed that the unsaturated hydrocarbons are less diffusible than the paraffin hydrocarbons. Later, Gilpin and Cram clearly demonstrated that when petroleum is allowed to diffuse through tubes packed with fuller's earth, the unsaturated hydrocarbons collect in the earth of lower sections of the tubes, while the paraffins tend to accumulate in the lightest fraction at the top of the tube. In the

[1] Z. angew. Chem., 1901, 889.

present investigation, these results have been fully confirmed. On pages 50 to 52 are given the bromine absorption values, and the percentages by volume absorbed by concentrated sulphuric acid of the various oils obtained from definite sections of a tube. These figures indicate conclusively that the amount of unsaturated hydrocarbons in the oils from the lower sections of the tube is much greater than the amount of these hydrocarbons in the lightest fractions at the top of the tube. Furthermore, the bromine absorption values for the oils of similar fractions of the first, second, and third fractionation, given on page 51, show that in the progress of the fractionation more and more of the unsaturated hydrocarbons are removed. Herr,[1] in Russia, has likewise observed that these hydrocarbons are less diffusible than the paraffins.

An interesting confirmation in nature of these experiments has been recently presented by Clifford Richardson and K. G. MacKenzie.[2] They found that a colorless natural naphtha from the Province of Santa Clara, Cuba, contained practically no unsaturated hydrocarbons, but was almost entirely a mixture of naphthenes and paraffins. Concentrated sulphuric acid absorbed but 0.76 per cent. by volume, while fuming sulphuric acid absorbed only 1.8 per cent. With the naphtha were obtained water and an emulsion of water, oil and clay. These investigators are of the opinion that the naphtha was "undoubtedly formed by the upward filtration of heavy petroleum through the clay stratum, similar to the fuller's earth filtrations of Gilpin and Cram, and the light naphtha in the upper part of the stratum was afterwards partly liberated by saline waters, the oil remaining in the clay forming with water the emulsion."[2]

A comparison of the proportions of unsaturated hydrocarbons in Ohio and Pennsylvania oils shows that the latter contains a much smaller percentage of these hydrocarbons. By assuming that the Pennsylvania oil diffused upward through such porous media as shales and limestones to its present location in the sandstones, it is possible to account for the

[1] Petroleum August, 1909.
[2] Am. J. Sci., May, 1910.

smaller amounts of the olefins in it on the basis of the experimental work described above. In its passage through the capillary interstices of the clays, limestone and sandstones, a fractionation, resulting in the removal of the unsaturated hydrocarbons, probably occurred. It is reasonable to conclude, therefore, that the variation in the content of unsaturated hydrocarbons between the Ohio, Texas, and California oils, on the one hand, and the Pennsylvania oil on the other, can be probably accounted for by assuming that the latter was subjected to capillary diffusion at some time in its career. That the light-colored naphthas occurring in various parts of the world were originally darker and heavier oils, and that their primitive character was changed by diffusion through media possessing the power of fractionation, seems very probable.

The behavior of the aromatic hydrocarbons, in particular benzene, in passing through fuller's earth, constitutes one of the subjects of this investigation. The results of this study, given in detail on pages 18 to 30, indicate clearly that benzene, like the olefins, tends to collect in the lower sections of a tube of fuller's earth through which the benzene, in solution, is allowed to diffuse. That the aromatic hydrocarbons in the natural oil behave in a similar manner has not yet been decided. The proportion of these hydrocarbons in the Illinois oil investigated was too small to enable us to determine accurately their amounts in the various fractions obtained by the capillary diffusion of the crude oil. The ordinary methods, such as nitration with a mixture of nitric acid and sulphuric acids, and sulphonation, employed for the quantitative determination of the aromatic hydrocarbons, could not be used in this work, owing to the fact that these reagents readily affect the unsaturated hydrocarbons as well. A study of the conduct of the aromatic hydrocarbons in the natural oil containing large amounts of them will be undertaken in the near future. It is probable, however, that the benzene and homologous compounds in crude petroleum behave like the unsaturated hydrocarbons.

The presence of larger amounts of aromatic hydrocarbons

in the Ohio than in the Pennsylvania petroleum, and still larger amounts in the California and Texas oils, seems to afford further evidence in favor of the view that the Pennsylvania oil has undergone much greater diffusion, and consequently greater fractionation, than any of the other oils.

The conduct of the sulphur compounds in petroleum in the process of diffusion is similar to that of the unsaturated hydrocarbons. On page 52, the percentages of sulphur present in the oils from different parts of the tube and different stages of fractionation are tabulated. One series of figures will be given to show the behavior of the sulphur compounds:

Lot 6.	Per cent. of sulphur.	
	First frac-tionation.	Third frac-tionation.
Fraction A	0.04	0.003
" B	0.05	. . .
" D	0.09	. . .
" E	0.16	. . .

It is clear from these figures that the sulphur compounds, like the unsaturated hydrocarbons, tend to collect in the lower sections of a layer of fuller's earth through which petroleum is allowed to diffuse.

In 1902, Clifford Richardson and E. C. Wallace,[1] in an investigation on the occurrence of free sulphur in Beaumont petroleum, passed this oil upward through the kaolin filter described by Dr. D. T. Day at the Petroleum Congress in Paris, in 1900, and obtained a distinct fractionation. The percentages of sulphur in the crude oil, and the oils obtained by this fractionation were determined. The results are given in the following table:

	Sp. gr. 25°/25°.	Per cent. sulphur.
Crude oil	0.9140	1.75
1st fraction	0.8775	0.70
2nd fraction	0.8986	0.91
3rd fraction	0.9038	1.04

It seems reasonable to assume from these results that the variations in the sulphur content between the Pennsylvania

[1] J. Soc. Chem. Ind., March, 1902.

and Ohio oils may be satisfactorily explained by the view that the former oil, as previously stated, diffused from other regions to its present location, and in its migration a large part of its original content of sulphur was removed. Further work upon this point will be undertaken in this laboratory.

No careful study of the behavior of the nitrogen and oxygen compounds in petroleum diffusing through a porous medium has as yet been undertaken. A careful investigation of this matter will be pursued in this laboratory later on. It is probable that such an investigation will show that the nitrogen compounds conduct themselves like the sulphur and unsaturated compounds.

The Object of this Investigation.

The present investigation was undertaken for the immediate purpose of studying the changes occurring in the crude Illinois oil when allowed to diffuse through fuller's earth. The more distant, but more fundamental, object was to gain further insight into the causes of the variations between the various oils of this country.

EXPERIMENTAL.

Preliminary Experiments.

The Relative Amounts of Oil Lost in Heated and Unheated Fuller's Earth.—Before the actual investigation of the Illinois oil was undertaken, experiments were made to determine the relative amounts of oil lost in heated and unheated fuller's earth.[1] In the work of Gilpin and Cram, the earth was always heated until geysers ceased to form, and then allowed to cool for several hours. The purpose of heating the earth was to obtain larger yields of oil, but towards the close of their investigation it became apparent that the amount of oil lost in unheated fuller's earth was not as large as they had supposed it to be. Since much time and labor is consumed in the process of heating and then cooling the earth, it seemed advisable to settle this point at the outset.

Apparatus.—The apparatus employed for this investiga-

[1] The fuller's earth employed in these investigations was generously supplied by the Atlantic Refining Company of Philadelphia.

tion was essentially the same as that used by Gilpin and Cram. *A, A, A, A* (Fig. I) are tin reservoirs made to hold somewhat more than a liter. The tin tubes *B, B, B, B,* 5.5 feet in length, and 1.25 inches in diameter, rest upon narrow tin supports placed upon the bottom of the reservoirs, and are connected with the branched glass tube *F* by suction tubing fitted with pinchcocks at *E, E, E, E.* The tube *F* is connected with the large tank *C,* which serves to maintain fairly constant pressures; *C* is in turn joined by the glass tube *D* to a manometer, and the latter connected with the Chapman pump. Any number of these tubes may be set up in series under the same diminished pressure.

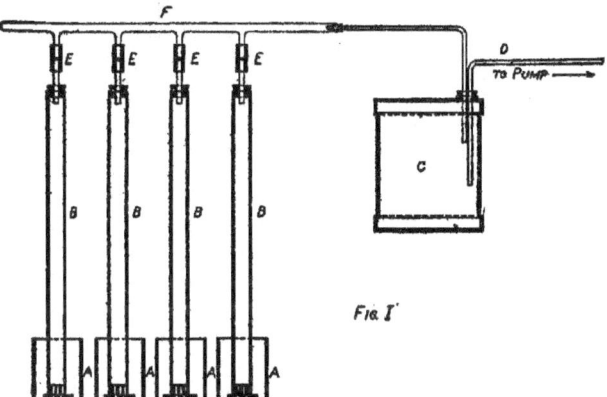

Fɪɢ. I'

After the tubes are closed at their lower ends with grooved corks covered with muslin to prevent the earth from sifting out, they are packed to the desired firmness with the fuller's earth. Each tube is then placed in its own reservoir, containing the oil to be fractionated. When they are connected to the branched tube *F,* the pressure in the system of tubes is reduced by the suction pump. The oil rises at first rapidly, then its diffusion gradually diminishes in power. When the reservoirs are almost exhausted, the tubes are disconnected and clamped, with the bottom ends up, above shorter tubes of the same diameter, into which the oil-laden earth is allowed to slide. These shorter tubes are made of two curved pieces, joined at the bottom by a cap, and held together at the top by a ring. The cylinders are opened by slipping off the ring and cap and removing one of the curved pieces, and the earth divided into the desired sections. When water is added in portions to the earth and the two mixed thoroughly, the oil is displaced and is drawn off in separate portions.

Six tubes packed with heated fuller's earth were set up alternately with six tubes filled with the unheated earth. Each tube was placed in its own reservoir containing 950 cc. of crude oil. The oil was allowed to diffuse upward through the tubes under diminished pressure. Sixteen hours elapsed before the oil in the reservoirs was exhausted. Since the tubes did not rest directly upon the bottom of the reservoirs, a small amount of oil remained, the volume of which was subtracted from the volume originally supplied. The earth from each tube was shaken into a bucket, and the oil recovered by displacement with water, as described above. The results of these experiments are given in the following table:

Table I.—Heated Fuller's Earth.

Tubes.	Weight of fuller's earth. Grams.	Oil absorbed by earth. cc.	Oil recovered. cc.	Oil lost. cc.	Per cent. oil lost cc.
1	1005	850	450	390	46
3	1000	792	460	332	41
5	1035	850	500	350	41
7	1070	865	450	415	48
9	1035	813	430	383	47
11	1045	885	530	355	41
Total,	....	5055	2830	2225	44

Unheated Fuller's Earth.

2	1075	917	585	332	36
4	1095	853	562	291	34
6	1065	840	500	340	42
8	1045	814	435	379	46
10	1035	873	510	363	41
12	1055	850	485	365	41
Total,	....	5147	3077	2070	40

The petroleum employed in the above experiments was a dark, green oil from Venango County, Pennsylvania, possessing a specific gravity of 0.810.

Since the Illinois oil, which was used in the fractionation proper, described later, differs materially from the Pennsylvania petroleum, further experiments were undertaken to determine the relative amounts of this oil retained by heated and unheated earth.

Ten tubes, five of which were packed as uniformly as possible with fuller's earth that had been heated until geysers ceased to form, and the other five with unheated earth, were placed in reservoirs, each containing 950 cc. of Illinois oil, specific gravity 0.8375. When the oil was entirely absorbed, the tubes were taken down and the oil-laden earth shaken into two breakable cylinders and divided into the following sections: A constitutes the section, 10 cm. in length, measured downward from the level to which the oil had ascended; B, the next 15 cm.; C, 20 cm.; D, 30 cm.; E, 35 cm.; the remainder of the earth to the bottom of the tube, designated as F, was entirely discarded.

The earth was then treated with separate portions of water. The oils displaced by the successive additions of water were collected separately and are designated in the table below as A_1, A_2, B_1, B_2, and so on; A_1 is the oil first displaced, A_2 the oil next expelled by further additions of water. The volumes and specific gravities of the recovered oils were determined. The results are given in the following table:

Table II.

Frac.	Heated fuller's earth.		Unheated fuller's earth.	
	Spec. grav.	Vol., cc.	Spec. grav.	Vol., cc.
A_1	0.8287	100	0.8320	72
A_2	. . . .	. . .	0.8352	22
B_1	0.8390	157	0.8405	184
B_2	0.8485	35	0.8451	124
C_1	0.8441	280	0.8443	270
C_2	0.8507	67	0.8495	147
D_1	0.8450	393	0.8483	368
D_2	0.8490	132	0.8517	210
E_1	0.8537	339	0.8500	360
E_2	0.8564	174	0.8569	185
		1701		1942

These results indicate that unheated fuller's earth retains no more oil than the heated earth. Although, in these experiments, the percentage of oil lost in the unheated is smaller than that lost in the heated earth, Gilpin and Cram, employing heated earth, recovered, in one test, 5,951 cc. from 9,070 cc., and, in another, 5,415 cc. from 8,915 cc., the amount of oil lost in the earth in the first test corresponding to 34 per cent., in the second to 39 per cent. It is clear, therefore, that there is no sufficient, if any, compensation for the time and labor spent in heating the earth. In the investigations that followed, therefore, the unheated fuller's earth was always used.

The Diffusion of Benzene in Solution through Fuller's Earth.— In order to deal more intelligently with the fractionation of the crude Illinois petroleum, it seemed advisable to study the behavior of the individual aromatic hydrocarbons, especially benzene, both alone and mixed with paraffin hydrocarbons, when allowed to diffuse upward through fuller's earth. Gilpin and Cram established the fact that the paraffin hydrocarbons tend to collect in the lightest fractions at the top of the tube. Their method consisted in distilling by heat six samples of oils of different specific gravities, each 300 cc. in volume, and collecting ten fractions between definite intervals. Five of these samples consisted of oil partly fractionated by fuller's earth, and the other of the crude oil. The specific gravity and viscosity of each fraction were determined; then to 30 cc., or to all there was where the amount was less than 30 cc., an equal volume of concentrated sulphuric acid (specific gravity 1.84) was added, and the two shaken in a machine for half an hour or longer. The volume of the oil unaffected by the acid was measured, and, by subtraction, the volume of oil absorbed was calculated. This latter volume represents only approximately the percentage of unsaturated hydrocarbons present in the oil, because sulphuric acid of this strength readily dissolves benzene when the two are thoroughly shaken.

In this investigation various solutions of benzene and a refined paraffin oil, boiling between 160° and 240°, and only slightly attacked by sulphuric acid, were made up and allowed to rise in tubes packed with unheated fuller's earth. The pressure in the system was reduced very little, because the liquid, under a greatly diminished pressure, rose too rapidly. About 24 hours elapsed before the oil in the reservoirs was exhausted.

The earth in each tube was shaken out and divided into six sections. Beginning at the uppermost point to which the oil had ascended, grade A consisted of the first 8 cm.; grade B of the next 8 cm.; grade C of 18 cm.; grade D of 30 cm.; grade E of 35 cm.; and, finally, grade F of the remainder of the earth, depending on the height to which the oil had ascended. This division is the same as that used by Gilpin and Cram. The oil in the earth was displaced by water and drawn off.

The specific gravity of each fraction was determined by means of the Mohr-Westphal balance at exactly 20°. The fourth decimal is not to be considered as strictly accurate, but gives a closer approximation to the truth than if it were entirely discarded.

The viscosity was determined by means of the viscosometer described by Ostwald and Luther and modified by Jones and Veazey.[1] The time taken for measured volumes of the oils to drain from the small bulb, whose capacity was 4.5 cc., was compared with the time required for a similar amount of water to run through. These values were substituted in the equation

$$y = y_0 \frac{TS}{T_0 S_0},$$

where y_0 = coefficient of viscosity of water. For this, 0.01002, the value obtained by Thorpe and Rodger,[2] was used.

t = time of flow of liquid under examination.

S = specific gravity, measured at 20°, of liquid under examination.

T_0 = time of flow of water.

S_0 = specific gravity of water. Since the balance was calibrated for water, at 20°, the value for S is unity.

y = coefficient of viscosity of oil under examination.

[1] Z. physik. Chem., 61, 351.
[2] Phil. Trans., A, 185, 397 (1894).

The amount of benzene present in each fraction was determined by shaking the oil with an excess of ordinary concentrated sulphuric acid (specific gravity 1.84) for periods of time varying from 30 to 60 minutes, until there was no further diminution in the volume of the oil.

The following experiments demonstrate the power of this acid to dissolve benzene, forming benzenesulphonic acid:

(1) Twenty-five cc. of benzene were shaken vigorously in a machine with 25 cc. of concentrated sulphuric acid (specific gravity 1.84) for 30 minutes. Amount of benzene dissolved, 7 cc., or 28 per cent.

(2) Twenty-five cc. were shaken for 30 minutes with 50 cc. of acid. Amount of benzene dissolved, 18 cc., or 72 per cent.

(3) Twenty-five cc. were shaken for 30 minutes with 75 cc. of acid. Amount of benzene dissolved, 25 cc., or 100 per cent.

The reagents usually employed for removing benzene are a mixture of fuming nitric and concentrated sulphuric acids. The work of Worstall,[1] Francis and Young,[2] and others, shows that such a mixture readily attacks the paraffin hydrocarbons, especially at higher temperatures, forming nitro derivatives, and also oxidizing them to a considerable extent. Furthermore, in working with this mixture the oil must be kept at a low temperature to prevent a violent reaction which results usually in the decomposition of the oil. In this work, therefore, in order to avoid the danger of attacking the paraffin hydrocarbons, and or the sake of convenience, concentrated sulphuric acid was used.

It seems advisable, at this point, to call attention to the fact that the power of ordinary concentrated sulphuric acid to remove benzene and homologous hydrocarbons has been generally overlooked. In order to determine the percentages of these hydrocarbons, it is customary to shake the oils to be analyzed with concentrated sulphuric acid, and then to nitrate the unaffected oil. It is assumed that the acid removes such substances as the unsaturated hydrocarbons, and does not attack the aromatic hydrocarbons. Thus, P.

[1] Amer. Chem. J., 20, 202; 21, 210.
[2] J. Chem. Soc., 1898, 928.

Poni,[1] in determining the presence and percentage of aromatic hydrocarbons in Roumanian petroleum, collected fractions between 35° and 70°, distilled under diminished pressure. These were *purified by shaking with sulphuric acid,* and each nitrated with a mixture of 1 part of nitric acid (specific gravity 1.52) and 2 parts sulphuric acid (specific gravity 1.8). The recovered oils were assumed to be paraffins and naphthenes, while the proportion of benzene and unsaturated hydrocarbons was calculated from the nitro products obtained. It is obvious from the results obtained in the present work that some of the benzene was removed in the process of purifying the fractions. The amount dissolved would depend upon the vigor of the shaking and its duration, as well as on the strength of the sulphuric acid. It is highly probable, therefore, that his percentage of benzene is too low.

In the study of the mixture of benzene and paraffin hydrocarbons, twenty-five cc. of each fraction, or the whole fraction when it was less than 25 cc., were shaken vigorously with three times their volume of concentrated sulphuric acid for 30 minutes. The amount unabsorbed was measured over the acid in a burette, after sufficient time was allowed for most of the oil that was mechanically held in suspension to rise. The oil was then reshaken with a little more acid 15 minutes longer, and the volume again read. In cases where the benzene was present only in small quantities one shaking was sufficient, in other cases it was repeated a second time.

The paraffin oil employed, specific gravity 0.797, was shaken several times with fresh portions of concentrated sulphuric acid until the acid was no longer colored, and only a slight diminution in volume occurred when a small sample of the oil was thoroughly shaken in a machine for some time with the acid. The oil was then washed with water and sodium hydroxide and dried over calcium chloride. The specific gravity decreased to 0.792.

When this oil was mixed with benzene in various proportions, and allowed to diffuse upward through fuller's earth, the following results, arranged in series, were obtained:

[1] Ann. Sci. Univ. Jassy, 1907, 192-202 (abstracted in J. Chem. Soc., 92, II, 883 (1907)).

Table III.

Series 1.—Oil alone. Specific gravity 0.792. Level of oil, 28 cm.

Grade.	Volume of oil, cc.	Specific gravity.	Viscosity.	Per cent. benzene.[2]
A	11	0.789	....	....
B	17	0.792	....	....
C	60	0.7912	0.0154	....
D	100	0.7915	0.0140	....
E	150	0.7913	0.0134	....
F	139	0.7915	0.0134	....

477[1]

Orig. vol., 778

Series 2.—90 per cent. oil (0.792)–10 per cent. benzene (0.8775). Specific gravity, 0.7983. Level of oil, 22 cm.

Grade.	Volume of oil, cc.	Specific gravity.	Viscosity.	Per cent. benzene.
A	11	0.787	....	10.0
B	16	0.7923	....	13.3
C	56	0.7935	0.0131	11.6
D	109	0.7943	0.0123	14.8
E	145	0.7957	0.0120	14.4
F	245	0.7955	0.0116	14.8

582

Orig. vol., 872

The results that are tabulated in the various series are ex-pressed diagrammatically in the following curves. The ordinates represent the different grades of oil, and the abscissas, the percentages of benzene and the specific gravities.

The final curve represents *in toto* the results of the experimental work upon the diffusion of benzene in solution through fuller's earth. The ordinates of this curve represent the percentages of benzene, and the abscissas, the various mixtures of benzene and oil that were allowed to diffuse through the earth.

[1] The original volumes of solution vary with each series, owing to the fact that more or less always remained behind in the reservoir below the level of the tin support. In Series 1, 2, 3, and 4, 950 cc. were supplied to each reservoir; in the rest of the series, each reservoir contained originally 1,000 cc.

[2] In this series the percentages of benzene are not given, because the paraffin oil alone was used.

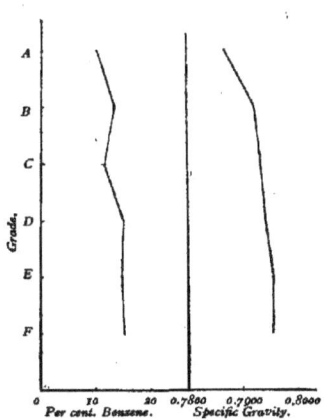

Per cent. Benzene.   Specific Gravity.
Fig. II.   Series 2.

Series 3.—80 per cent. oil (0.792)–20 per cent. benzene (0.8775).
Specific gravity, 0.806.   Level of oil, 25 cm.

Grade.	Volume of oil, cc.	Specific gravity.	Viscosity.	Per cent. benzene.
A	25	0.7948	0.0147	15.3
B	35	0.7981	0.0130	16.0
C	78	0.8017	0.0117	22.4
D	126	0.8005	0.0105	21.6
E	166	0.801	0.0107	22.4
F	146	0.798	0.0110	20.8

	576
Orig. vol.,	892

Series 4.—75 per cent. oil (0.792)–25 per cent. benzene (0.8775). Specific gravity, 0.810.   Level of oil, 33 cm.

Grade.	Volume of oil, cc.	Specific gravity.	Viscosity.[1]	Per cent. benzene.
A	16	0.800	....	22.0
B	35	0.803	0.0129	23.3
C	74	0.8077	0.0126	24.0
D	128	0.805	0.0114	24.0
E	152	0.8068	0.0102	26.0
F	120	0.8065	0.0105	28.0
	———			
	525			
Orig. vol.,	655			

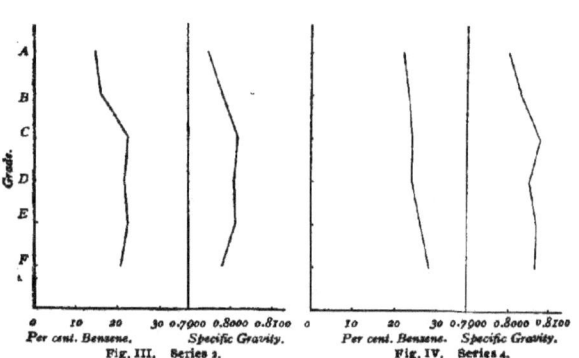

0   10   20   30 0.7900 0.8000 0.8100   0   10   20   30 0.7900 0.8000 0.8100
Per cent. Benzene.   Specific Gravity.   Per cent. Benzene.   Specific Gravity.
Fig. III.   Series 3.   Fig. IV.   Series 4.

[1] The viscosities of Grades *A* and *B* in a few of the tables are not given, because, in these series, which were the first to be made, the decision to determine the viscosities was reached only after the fractions had been treated with acid. Since *A* and *B* were small, all the oil was used up in this treatment.

Series 5.—75 per cent. oil (0.794)[1]–25 per cent. benzene (0.8775). Specific gravity, 0.8115. Level of oil, 24 cm.

Grade.	Volume of oil, cc.	Specific gravity.	Viscosity.	Per cent. benzene.
A	25	0.7942	0.0123	14.0
B	28	0.8048	0.0104	21.2
C	70	0.8105	0.0094	31.2
D	140	0.8100	0.0094	27.6
E	172	0.8100	0.0094	32.0
F	144	0.8093	0.0095	27.6

	579
Orig. vol.,	875

Series 6.—75 per cent. oil (0.792)–25 per cent. benzene (0.8775). Specific gravity, 0.8083. Level of oil, 27 cm.

Grade.	Volume of oil, cc.	Specific gravity.	Viscosity.	Per cent. benzene.
A	22	0.7995	0.0106	17.5
B	32	0.8055	0.0099	24.4
C	82	0.8052	0.0100	24.0
D	155	0.8085	0.0093	28.8
E	190	0.8085	0.0093	31.2
F	93	0.8063	0.0096	28.8

	574
Orig. vol.,	923

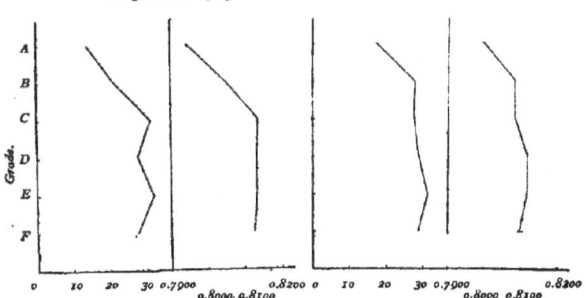

Per cent. Benzene. Specific Gravity. Per cent. Benzene. Specific Gravity.
Fig. V. Series 5. Fig. VI. Series 6.

[1] In Series 5, 8, 9 and 10 the specific gravity of the refined oil is 0.794. Since the quantity of oil of specific gravity 0.792 was not sufficient for all the series, a second quantity was prepared which had the specific gravity 0.794. This oil was used in the above-mentioned series.

Series 7.—59.5 per cent. oil (0.792)–40.5 per cent. benzene (0.8775). Specific gravity, 0.8223. Level of oil, 9 cm.

Grade.	Volume of oil, cc.	Specific gravity.	Viscosity.	Per cent. benzene.
A	9[1]	....	....	....
B	15	0.8069	....	14.0
C	48	0.816	0.0103	22.4
D	96	0.8182	0.0086	31.2
E	160	0.820	0.0082	31.6
F	255	0.8185	0.0083	29.6

583
Orig. vol., 922

Series 8.—50 per cent. oil (0.794)–50 per cent. benzene (0.8775). Specific gravity, 0.8295. Level of oil, 17 cm.

Grade.	Volume of oil, cc.	Specific gravity.	Viscosity.	Per cent. benzene.
A	22	0.8122	....	24.5
B	32	0.819	....	28.4
C	78	0.8287	0.0077	44.8
D	111	0.8275	0.0077	47.6
E	155	0.827	0.0077	39.2
F	192	0.8256	0.0079	36.4

590
Orig. vol., 960

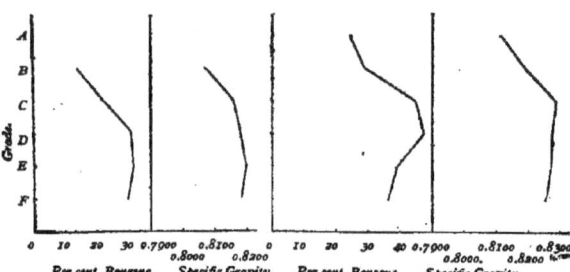

Per cent. Benzene. Specific Gravity.   Per cent. Benzene.   Specific Gravity.
Fig. VII.   Series 7.   Fig. VIII.   Series 8.

[1] In Series 7 the volume of Grade *A* recovered was so small that no measurements could be made.

Series 9.—50 per cent. oil (o.794)–50 per cent. benzene (o.8775).
Specific gravity, o.8315. Level of oil, 18 cm.

Grade.	Volume of oil, cc.	Specific gravity.	Viscosity.	Per cent. benzene.
A	18	0.816	0.0091	26.0
B	24	0.8210	0.0085	34.5
C	76	0.8275	0.0078	47.6
D	136	0.8283	0.0077	50.0
E	174	0.8293	0.0076	49.2
F	144	0.8277	0.0078	40.0

572
Orig. vol., 923

Series 10.—50 per cent. oil (o.794)–50 per cent. benzene
(o.8775). Specific gravity, o.8295. Level of oil, 16 cm.

Grade.	Volume of oil, cc.	Specific gravity.	Viscosity.	Per cent. benzene.
A	31	0.8135	0.0097	31.6
B	45	0.8251	0.0081	43.6
C	85	0.8290	0.0076	46.4
D	140	0.8280	0.0077	47.6
E	175	0.8285	0.0076	49.6
F	137	0.8272	0.0076	50.0

613
Orig. vol., 972

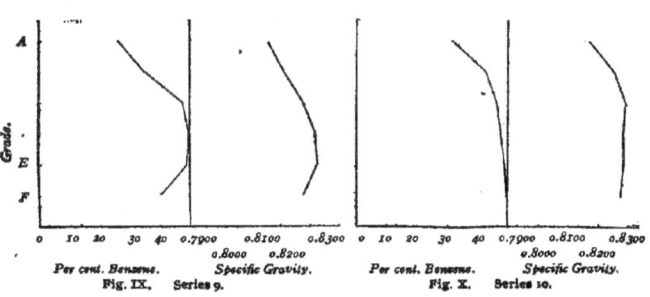

Fig. IX.   Series 9.   Fig. X.   Series 10.

Series 11.—75 per cent. crude oil (0.810)–25 per cent. benzene (0.8775). Specific gravity, 0.8312. Level of oil, 18 cm.

Grade.	Volume of oil, cc.	Specific gravity.	Viscosity.	Per cent. benzene.[1]
A	12	0.8255	0.0445	....
B	22	0.8268	0.0423	....
C	52	0.8280	0.0300	....
D	76	0.8290	0.0298	....
E	140	0.8300	0.0263	....
F	186	0.8320	0.0276	....

488

Orig. vol., 890

Series 12.—Benzene alone (0.8775). Level of oil, 33 cm.

Grade.	Volume of oil, cc	Specific gravity.	Viscosity.	Per cent. benzene.
A	16	0.8765	....	....
B	15	0.877	....	....
C	68	0.878	0.0066	....
D	128	0.8778	0.0066	....
E	157	0.8775	0.0066	....
F	89	0.8771	0.0066	....

473

Orig. vol., 888

An examination of these figures shows conclusively that benzene tends to collect in the lower portions of the tube. The specific gravities and viscosities confirm the results obtained by determining the percentages of benzene present by removing the benzene with concentrated sulphuric acid. The specific gravities of Grades F to C run very close together, and are all much greater than those of Grades A and B. Since benzene possesses a high specific gravity (in this work the specimen had a specific gravity of 0.8775), the larger value for the lower grades indicates the presence of larger amounts of benzene. The specific gravity of the paraffin oil was only 0.792, showing that the higher specific gravities were due to larger percentages of benzene. Further, since the viscosity

---

[1] The percentages of benzene in Series 11, in which crude oil was employed, are not recorded, because, owing to the formation of heavy black emulsions, the loss in volume could not be determined with any degree of accuracy.

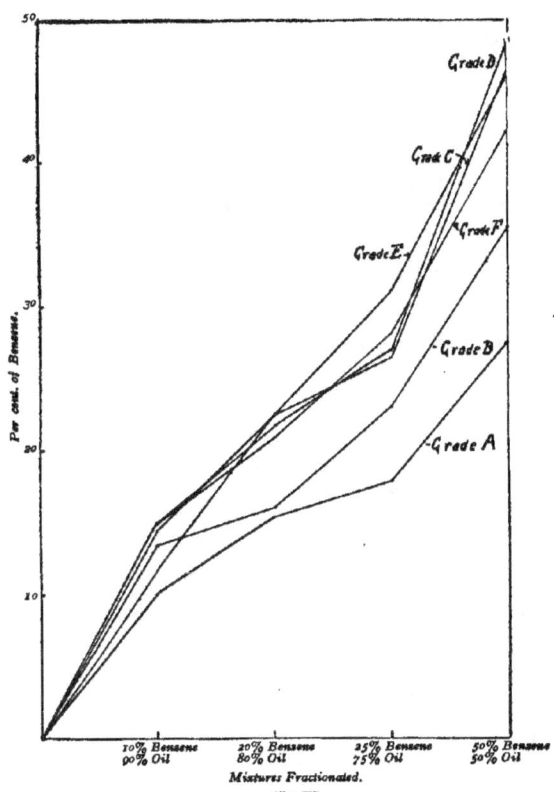

Fig. XI.

of the benzene used was 0.0066, and that of the paraffin oil about 0.0150, the viscosities of those fractions containing higher percentages of benzene, we should expect, ought to be much smaller than those containing less benzene. The results show that the viscosities of the Grades F to C are much smaller than those of A and B.

It will be observed that the maximum in specific gravity is reached, not at F, as might be expected in the fractionation of the crude oil, but between C and D. Between B and C there is a marked decrease. This sudden break is found also in the viscosities, and in the percentages of benzene. While the sharp breaks in the curves represent the marked change in the proportion of benzene and the height to which it rises in the tube, no satisfactory explanation has yet been obtained as to why it should occur at these points. This action will be studied more carefully later.

In order to determine the degree of exactness of the percentages of benzene obtained, known amounts of benzene were added to the oil until the specific gravity corresponded closely to that obtained by fractionation. The amount of benzene thus added and the amount actually removed by the acid agree very closely, as the following results show:

Benzene in 25 cc. of mixture.	Specific gravity.	Benzene found in the grades of Series 8. cc.		Specific gravity.
7.3	0.8143	Grade A	7.9	0.8135
9.4	0.8213	" B	10.9	0.8251
11.1	0.8274	" F	12.5	0.8272
11.3	0.8287	" E	12.4	0.8287
11.9	0.8293	" C	11.6	0.8290

The variations in the specific gravities of the mixtures and those of the grade A–F are due to the fact that in the latter series some fractionation had taken place, and therefore the paraffin oils mixed with the benzene were not identical with those mixed with the benzene in the series of prepared mixtures, as the paraffin oil used was not an individual substance but a mixture.

The Fractionation of Crude Petroleum.

The petroleum employed for the fractionation was an oil obtained from the E. E. Newlin farm, 2.5 miles west of Robinson, Crawford County, Illinois. The specific gravity of the oil was 0.8375 at 20°; its color was dark brown.

The fractionation of the oil was effected by upward diffusion through tubes packed with fuller's earth. In order to shorten the time required for the oil to diffuse by capillarity to the upper parts of the tube, the fine interstices and pores of the earth were evacuated by applying diminished pressure at the top of the tube. By this aid, the time required for the oil to reach the top of a tube was reduced from several weeks to one or two days.

The apparatus employed is the same as that described on page 14.

The tin tubes were packed as uniformly as possible by introducing definite amounts of earth, and ramming solidly with rods tipped with rubber stoppers. The degree of compactness depended upon the kind of oil to be used. For the crude oil, about one and one-half feet of the tube was filled at a time, and the earth packed as firmly as possible; for the lighter oils, one foot of the tube was filled at a time; for the oils heavier than the crude, between two and three feet of the tube were filled at one time.

The tubes were then placed individually in reservoirs containing 950 cc. of the crude oil, after which diminished pressure was applied at the top of the tubes. The oil rose rapidly at first, then diffused more and more slowly as the tops of the tubes were approached. When the oil in the reservoirs was completely exhausted, the tubes were disconnected from the blanched glass tube F (see Fig. I) and the oil-laden earth shaken into two breakable cylinders. For the various fractions, the following divisions of the earth were made: Fraction A constituted the first 10 cm., measured downward from the level to which the oil had ascended; fraction B, the next 15 cm.; C, 20 cm.; D, 30 cm.; E, 35 cm., and F, the remainder to the bottom of the tube. In the first fractionation up to

Lot 28, fraction F was dicarded; from Lot 28 to the end of the first fractionation, E and F were collected together.

After thus dividing the earth, the various portions were placed in separate receptacles and treated with water. After each addition of water the two were thoroughly mixed. The earth, when the oil first appears, is granular; as more water is added, liberating more oil, the earth becomes muddy, and when as much oil as possible has been expelled by the water, the earth has the consistency of glue.

The portions of oil liberated by successive additions of water were collected separately. As Gilpin and Cram[1] pointed out, the oil that is first expelled, if not very small in volume as compared with the oils succeeding, possesses a lower specific gravity than the oil liberated by further additions of water; the latter, in turn, is lighter than the next succeeding oil. The oil that is liberated last, therefore, possesses a higher specific gravity than any of the oils preceding it. Sometimes, however, the specific gravity remains constant after the second or third extraction. This fractionation, by means of water, was combined with the fractionation effected by the fuller's earth. In the tables that follow, A_1 is the oil first liberated, A_2 the oil next liberated; in the lower fractions, $i.\ e.$, C, D, E, three and sometimes four extractions were made before all the oil that could possibly be liberated by water was recovered.

The specific gravity of the oils was determined by means of the Mohr-Westphal balance. As mentioned before, the fourth decimal is not to be considered as rigidly accurate, but it gives a closer approximation to the truth than if it were entirely discarded. The temperature at which the specific gravity was measured was exactly 20°.

[1] Amer. Chem. Jour., 40, 495 (1908).

Table IV.—The First Fractionation.

Lot. No. of tubes.	1 15		2 5		3 10			
Hours req.	18—14 tubes 23—1 tube		16		17—8 tubes		45—2 tubes	
Frac.	Spec. grav.	Vol., cc.	Spec. grav.	Vol., cc.	Spec. grav.	Vol., cc.	Spec. grav.	Vol., cc.
A_1	0.8250	312	0.8285	73	0.8223	138	0.8233	50
A_2	0.8287	90	0.8310	59	0.8270	54	....	...
B_1	0.8367	485	0.8370	218	0.8372	258	0.8405	130
B_2	0.8392	250	0.8408	78	0.8400	200	....	...
C_1	0.8413	828	0.8440	272	0.8442	290	0.8505	120
C_2	0.8460	228	0.8442	136	0.8455	235	0.8535	65
C_3	0.8488	126	....	...	0.8480	148	....	...
D_1	0.8470	1014	0.8430	313	0.8488	538	0.8546	235
D_2	0.8495	375	0.8464	150	0.8500	295	0.8619	30
D_3	0.8514	200	0.8500	112	0.8540	115	....	...
D_4	0.8555	172	....	...	....	...	....	...
E_1	0.8527	720	0.8475	285	0.8537	380	0.8615	172
E_2	0.8540	430	0.8509	135	0.8550	245	....	...
E_3	0.8570	400	0.8540	118	0.8580	170	....	...

Lot. No. of tubes.	4 10		5 8		6 10[3]			
Hours req.	16		17—7 tubes 24—1 tube		17—1 tube[4] 40—3 tubes 96—1 tube		17—3 tubes 40—1 tube 150—1 tube	
Frac.	Spec. grav.	Vol., cc.	Spec. grav.	Vol., cc.	Spec. grav.	Vol., cc.	Spec. grav.	Vol., cc.
A_1	0.8295	170	0.8313	130	0.8320	72[5]	0.8287	85[1]
A_2	0.8315	100	0.8357	56	0.8352	22	....	...
B_1	0.8375	327	0.8392	358	0.8405	184	0.8490	134
B_2	0.8413	250	0.8453	92	0.8451	124	0.8485	35
C_1	0.8418	505	0.8419	425	0.8443	270	0.8441	218
C_2	0.8442	223	0.8439	138	0.8495	147	0.8507	67
C_3	0.8495	74	0.8465	130	....	...	....	...
D_1	0.8449	495	0.8454	640	0.8483	368	0.8450	302
D_2	0.8455	328	0.8500	167	0.8517	210	0.8490	132
D_3	0.8490	260	0.8509	195	....	...	....	...
E_1	0.8500	545	0.8495	575	0.8500	360	0.8537	215
E_2	0.8510	295	0.8513	185	0.8569	185	0.8564	174
E_3	0.8567	170	0.8555	130	....	...	....	...

[1] Chapman pump was run day and night. Manometer indicated pressures ranging from 30 to 80 mm.

[2] In lots 1 to 5, 1000 cc. of crude oil were supplied to each tube.

[3] Beginning with lot 6, 950 cc. of crude oil were supplied to each tube.

[4] The pressure in the tubes was diminished intermittently.

[5] See page 17.

Frac.	Lot 7 (9 tubes) 20—7 tubes		20—1 tube / 24—1 tube		Lot 8 (10 tubes) 19—8 tubes / 22—2 tubes		Lot 9 (10 tubes) 24—2 tubes / 40—8 tubes	
	Spec. grav.	Vol., cc.	Spec. grav.	Vol., cc.	Spec. grav.	Vol., cc.	Spec. grav.	Vol., cc.
A_1	0.8325	66	0.8175	45	0.8364	88	0.8215	145
A_2	0.8356	30	....	...	0.8365	64	0.8234	90
B_1	0.8395	164	0.8333	110	0.8400	215	0.8330	397
B_2	0.8418	140	....	...	0.8420	240	0.8350	155
B_3	....	...	....	...	....	...	0.8400	87
C_1	0.8408	475	0.8417	132	0.8445	368	0.8415	350
C_2	0.8468	123	0.8500	22	0.8467	225	0.6436	255
C_3	....	...	....	...	0.8495	82	0.8480	160[1]
D_1	0.8449	500	0.8468	110	0.8465	460	0.8485	507
D_2	0.8487	270	0.8498	106	0.8478	260	0.8495	280
D_3	....	...	....	...	0.8500	260	0.8545	247
E_1	0.8500	483	0.8533	228	0.8490	450	0.8548	313
E_2	0.8524	318	....	...	0.8495	354	0.8550	275
E_3	....	...	....	...	0.8521	233	0.8580	375

Frac.	Lot 10 (8 tubes) 14 hrs.		Lot 11 (10 tubes) 17 hrs.		Lot 12 (9 tubes) 42 hrs.		Lot 13 (10 tubes) 24—8 tubes / 40—2 tubes	
	Spec. grav.	Vol., cc.	Spec. grav.	Vol., cc.	Spec. grav.	Vol., cc.	Spec. grav.	Vol., cc.
A_1	0.8273	130	0.8258	215	0.8325	125	0.8323	122
A_2	0.8288	75	0.8318	70	0.8345	87	0.8352	96
B_1	0.8395	220	0.8370	340	0.8430	235	0.8438	245
B_2	0.8418	160	0.8480	180	0.8467	120	0.8470	180
C_1	0.8423	240	0.8422	488	0.8470	278	0.8464	317
C_2	0.8440	195	0.8450	205	0.8487	288	0.8505	235
C_3	0.8500	150	....	...	....	...	....	...
D_1	0.8460	410	0.8465	565	0.8495	452	0.8500	312
D_2	0.8475	210	0.8490	310	0.8522	305	0.8492	375
D_3	0.8500	348	0.8530	187	....	...	0.8518	150
E_1	0.8532	320	0.8510	297	0.8505	475	0.8505	450
E_2	0.8535	282	0.8520	405	0.8533	490	0.8489	395
E_3	0.8550	215	0.8533	155	....	...	0.8518	180

[1] Several cubic centimeters of this fraction were mixed, accidentally, with fraction E_3.

Lot.	14		15				16	
No. of tubes.	5		6				15	
Hours req.	24[1]		26— 3 tubes		26— 3 tubes		40— 11 tubes / 64— 4 tubes	
Frac.	Spec. grav.	Vol., cc.	Spec. grav.	Vol., cc.	Spec. grav.	Vol., cc.	Spec. grav.	Vol., cc.
A_1	0.8355	132	0.8381	60	0.8305	73	0.8370	200
A_2	....	...	....	...	....	...	0.8357	108
B_1	0.8470	236	0.8487	94	0.8452	143	0.8449	490
B_2	....	...	....	...	....	...	0.8445	226
C_1	0.8565	98	0.8430	110	0.8465	138	0.8475	635
C_2	0.8560	150	0.8480	57	0.8509	88	0.8509	235
C_3	....	...	....	...	....	...	0.8562	90
D_1	0.8523	170	0.8475	212	0.8505	158	0.8540	825
D_2	0.8550	205	0.8517	104	0.8522	178	0.8530	495
D_3	....	...	....	...	....	...	0.8575	150
E_1	0.8540	150	0.8467	184	0.8561	192	0.8538	775
E_2	0.8532	325	0.8502	152	0.8585	140	0.8562	620
E_3	....	...	....	...	....	...	0.8595	205

Lot.	17		18		19		20	
No. of tubes.	9		8		10		10	
Hours req.	40		24— 5 tubes / 48— 2 tubes / 64— 1 tube		40— 8 tubes / 64— 2 tubes		20— 6 tubes / 30— 4 tubes	
Frac.	Spec. grav.	Vol., cc.	Spec. grav.	Vol., cc.	Spec. grav.	Vol., cc.	Spec. grav.	Vol., cc.
A	0.8258	225	0.8322	112	0.8320	146	0.8281	236
B	0.8432	452	0.8435	335	0.8438	385	0.8413	518
C_1	0.8480	450	0.8495	250	0.8480	300	0.8450	350
C_2	0.8488	168	0.8500	250	0.8472	315	0.8495	300
D_1	0.8530	520	0.8530	320	0.8509	422	0.8508	325
D_2	0.8550	350	0.8540	350	0.8535	355	0.8538	460
E_1	0.8585	385	0.8547	90[2]	0.8492	580	0.8513	445
E_2	0.8598	460	0.8526	640	0.8560	415	0.8540	550

[1] When the pressure in the tubes was diminished, the oil rose rapidly, and in a short time, the reservoirs were nearly two-thirds exhausted. The pump was stopped, and the remainder of the oil allowed to diffuse during the night under normal pressure.

[2] This irregularity, i. e., the liberation of oil with a specific gravity higher than those of the oils immediately following, is observed when an amount of water is added sufficient to replace a very small amount of oil for the first fraction.

Lot.	21		22		23		24	
No. of tubes	10		10		10		10	
Hours[1] req.	24— 6,tubes 40— 2 tubes 64— 2 tubes		40— 6 tubes 64— 4 tubes		48— 5 tubes 52— 5 tubes		40— 4 tubes 64— 6 tubes	
Frac.	Spec. grav.	Vol., cc.	Spec. grav.	Vol., cc.	Spec. grav.	Vol., cc.	Spec. grav.	Vol., cc.
A	0.8275	245	0.8281	210	0.8241	330	0.8250	287
B	0.8410	615	0.8405	508	0.8395	615	0.8408	535
C_1	0.8452	520	0.8459	265	0.8448	420	0.8463	475
C_2	0.8488	226	0.8472	410	0.8470	305	0.8505	186
D_1	0.8512	533	0.8505	435	0.8533	400	0.8540	525
D_2	0.8535	415	0.8523	450	0.8541	465	0.8540	360
E_1	0.8557	375	0.8615	385	0.8650	305	0.8623	393
E_2	0.8625	282	0.8585	365	0.8624	350	0.8645	335

Lot.	25		26		27		28	
No. of tubes	9		10		10		10	
Hours[4] req.	48— 8 tubes 72— 1 tube		17— 2 tubes 24— 4 tubes 41— 4 tubes		17— 4 tubes 29— 6 tubes		24— 7 tubes 28— 3 tubes	
Frac.	Spec. grav.	Vol., cc.	Spec. grav.	Vol., cc.	Spec. grav.	Vol., cc.	Spec. grav.	Vol., cc.
A	0.8270	225	0.8284	315	0.8312	230	0.8333	240
B	0.8425	410	0.8422	550	0.8440	470	0.8440	410
C_1	0.8495	75[2]	0.8473	520	0.8460	400	0.8458	415
C_2	0.8492	250	0.8508	178	0.8478	232	0.8500	177
D_1	0.8509	320	0.8515	600	0.8482	435	0.8470	387
D_2	0.8510	480	0.8540	230	0.8500	420	0.8498	400
E_1	0.8556	335	0.8559	490	0.8520	465	0.8492	690[3]
E_2	0.8570	395	0.8586	135	0.8565	335	0.8505	600

Lot.	29		30		31		32	
No. of tubes	10		15		10		15	
Hours[4] req.	18— 5 tubes 40— 5 tubes		20— 7 tubes 41— 6 tubes 63— 2 tubes		44— 4 tubes 89— 6 tubes		40— 7 tubes 89— 4 tubes 103— 4 tubes	
Frac.	Spec. grav.	Vol., cc.	Spec. grav.	Vol., cc.	Spec. grav.	Vol., cc.	Spec. grav.	Vol., cc.
A	0.8262	300	0.8348	335	0.8292	245	0.8270	445
B	0.8395	505	0.8468	630	0.8439	576	0.8423	726
C_1	0.8463	390	0.8490	560	0.8495	465	0.8500	730
C_2	0.8488	270	0.8505	277	0.8523	205	0.8500	220
D_1	0.8520	510	0.8485	750	0.8517	670	0.8545	750
D_2	0.8543	290	0.8502	540	0.8552	210	0.8543	540
EF_1	0.8550	417	0.8520	1125	0.8555	805	0.8580	870
EF_2	0.8559	645	0.8528	880	0.8610	360	0.8598	910
		3327		5097		3536		5191

[1] Pressure in the tubes was diminished intermittently.
[2] Some oil of this fraction was lost.
[3] Beginning with lot 28, fractions E and F were collected together.
[4] Pressure in tubes was diminished intermittently.

Lot.	33		34		35	
No. of tubes.	10		10		9	
Hours¹ req.	41— 4 tubes		44— 6 tubes		48— 6 tubes	
	65— 4 tubes		68— 4 tubes		72— 3 tubes	
	89— 2 tubes					
Frac.	Spec. grav.	Vol., cc.	Spec. grav.	Vol., cc.	Spec. grav.	Vol., cc.
A	0.8330	290	0.8355	320	0.8380	235
B_1	0.8440	365	0.8475	525	0.8460	452
B_2	0.8462	165	....	...	....	.:.
C_1	0.8502	500	0.8508	470	0.8508	345
C_2	0.8540	160	0.8543	190	0.8525	245
D_1	0.8555	655	0.8575	530	0.8549	580
D_2	0.8562	250	0.8585	325	0.8573	335
EF_1	0.8575	735	0.8535	895	0.8557	645
EF_2	0.8585	480	0.8555	405	0.8570	492
		3600		3660		3329

Observations on the First Fractionation.

Specific Gravity.—The range of the specific gravity extended from 0.8175, the value for Fraction A_1 of Lot 7, to 0.8650, the value for Fraction E_1 of Lot 13. The value for the crude oil itself was 0.8375. The limits of the specific gravities of the individual lots averaged from 0.820 to 0.860. The specific gravity decreases gradually from E to B, but between B and A, the decrease, in most of the lots, is much greater than between any two consecutive lower fractions. This marked change was also observed in the study of the diffusion of benzene in solution. A detailed investigation into the cause of this sudden divergence will be undertaken in the near future.

Color.—The colors of the fractions obtained extended from green to black. The lighter oils possessed a beautiful green fluorescent color, which shaded gradually to brown, and then to the deep black of the heavier oils.

Odor.—The unpleasant odor of the crude petroleum disappeared almost entirely in the oils of Fraction A and B; but the other fractions still possessed to a greater or less extent the odor of the natural oil.

The Volume of Oil Retained by the Fuller's Earth.—The amount of oil retained by the earth averaged about 55 per

cent. of the amount supplied. In the first fractionation of
the crude Pennsylvania oil, specific gravity 0.810, Gilpin
and Cram found that approximately 40 per cent. of the oil
was retained by the earth. It is evident, therefore, that the
amount of oil remaining in the earth depends chiefly upon the
character of the oil. The Pennsylvania petroleum contains
a much smaller percentage of unsaturated hydrocarbons,
sulphur, and asphaltic substances than the Illinois oil employed in this investigation. Since the fuller's earth, as will
be shown later, readily removes these substances in the process
of fractionation, the large percentage of Illinois oil retained
by the earth is thus clearly explained. It is safe to conclude
that if the heavy Texas or California oil were allowed to diffuse
through fuller's earth, the amount of oil retained would ex-
ceed the amounts of either of the above-mentioned oils lost
in the earth.

The Second Fractionation.

The products obtained from the first fractionation were
united according to the following arrangement:

Lot.	Specific gravity of the oils united.	Specific gravity of mixture.
36	0.8250–0.8350	0.8293
37	0.8350–0.8400	0.8390
38	0.8400–0.8450	0.8433
39	0.8400–0.8450	0.8433
40	0.8450–0.8500	0.8490
41	" "	"
42	" "	"
43		
44	0.8500–0.8600	0.8543
45	" "	"
46	" "	"
47	" "	"
48	" "	"
49	" "	"
50	" "	"

The oils thus combined were subjected to chilling and filtration for the purpose of removing as much dissolved paraffin

as possible. The procedure was as follows: The oils were first chilled at temperatures ranging from $0°$ to $10°$ and then filtered through plaited filter papers. When the oil ceased to drip from the funnel, the residue upon the filter paper was placed in a larger filter press, and the remaining oil separated by pressure from the paraffin. The filter press was simple in construction. A piston, fitted closely in an iron cylinder, was gradually forced down upon the oil-laden paraffin, which rested upon a membrane of cotton duck fastened between perforated tin supports. The retained oil was forced through the membrane and was collected from the outlet below. The lighter oils deposited very little paraffin; from the heavier ones somewhat more paraffin was separated. Owing to the high viscosity of the heavier oils, the filtration proceeded very slowly. Since too much time was consumed in this process, the paraffin of some of the oils of Fraction E was not removed. A slight change in specific gravity occurred in the oils from which the paraffin was removed.

The final specific gravities of the united oils were as follows:

Lot.	Specific gravity.	
36	0.8305	Paraffin removed.
37	0.8415	" "
38	0.8433	Paraffin not removed.
39	0.8455	Paraffin removed.
40	0.8515	" "
41	0.8515	" "
42	0.8515	" "
43	0.8540	" "
44	0.8543	Paraffin not removed.
45	0.8543	" " "
46	0.8543	" " "
47	0.8543	" " "
48	0.8543	" " "
49	0.8557	Paraffin removed.
50	0.8557	" "

When these oils were again allowed to diffuse upward through fuller's earth, the following fractionation was obtained:

Table V.—The Second Fractionation.

Lot. No. of tubes.	36		37		38		39	
No. of tubes.	5		4		8		8	
Hours[1] req.	44—3 tubes 48—2 tubes		51		48—7 tubes 64—1 tube		29—4 tubes 45—3 tubes 64—1 tube	
Frac.	Spec. grav.	Vol., cc.	Spec. grav.	Vol., cc.	Spec. grav.	Vol., cc.	Spec. grav.	Vol., cc.
A	0.8272	160	0.8292	135	0.8331	180	0.8290	255
B_1	0.8315	216	0.8421	215	0.8447	175	0.8432	355
B_2	0.8331	58	....	...	0.8455	210	0.8458	110
C_1	0.8334	350	0.8467	295	0.8490	305	0.8492	455
C_2	0.8355	85	....	...	0.8505	175	0.8513	180
D_1	0.8330	360	0.8468	340	0.8492	400	0.8505	740
D_2	0.8339	320	0.8485	·152	0.8509	295	0.8527	275
EF_1	0.8347	720	0.8480	535	0.8508	710	0.8546	1166
EF_2	0.8356	320	0.8489	215	0.8518	355	0.8560	350
		2589		1887		3886		2805

Lot. No. of tubes.	40		41		42		43	
No. of tubes.	9		5		5		4	
Hours req.	48—5 tubes 72—4 tubes		40		69		10 days—2 tubes 17 days—2 tubes	
Frac.	Spec. grav.	Vol., cc.	Spec. grav.	Vol., cc.	Spec. grav.	Vol., cc.	Spec. grav.	Vol., cc.
A	0.8305	380	0.8316	235	0.8325	210	0.8435	65
B_1	0.8438	515	0.8460	290	0.8487	265	0.8546	115
B_2	0.8453	155	0.8480	65	0.8515	54	....	...
C_1	0.8518	600	0.8523	375	0.8540	335	0.8575	200
C_2	0.8539	170	0.8540	100	0.8567	56	....	...
D_1	0.8550	685	0.8558	470	0.8572	420	0.8605	220
D_2	0.8560	330	0.8571	110	0.8582	175	0.8640	50
EF_1	0.8605	780	0.8620	580	0.8640	675	0.8650	225
EF_2	0.8620	600	0.8622	320	0.8650	200	0.8615	78
		4215		2545		2420		953

[1] In this series, as well as those following, the pressure in the tubes was diminished intermittently.

Frac.	Lot. 44 No. of tubes 3 Hours req. 48—2 tubes 96—1 tube Spec. grav.	Vol. cc.	Lot. 45 No. of tubes 5 Hours req. 66 Spec. grav.	Vol. cc.	Lot. 46 No. of tubes 5 Hours req. 93 Spec. grav.	Vol. cc.	Lot. 47 No. of tubes 5 Hours req. 13 days[1] Spec. grav.	Vol. cc.
A	0.8330	85	0.8362	170	0.8332	210	0.8340	145
B_1	0.8505	175	0.8510	210	0.8480	260	0.8500	275
B_2	....	...	0.8522	80	0.8505	50	....	...
C_1	0.8582	155	0.8562	265	0.8554	300	0.8553	320
C_2	0.8605	65	0.8585	50	0.8567	95	0.8576	50
D_1	0.8605	195	0.8567	425	0.8600	370	0.8595	430
D_2	0.8620	120	0.8580	100	0.8613	120	0.8618	70
EF_1	0.8672	240	0.8659	615	0.8666	610	0.8665	330
EF_2	0.8680	175	0.8670	150	0.8680	130	0.8670	215
		1210		2065		2145		1835

Frac.	Lot. 48 No. of tubes 5 Hours req. 14 days[2] Spec. grav.	Vol. cc.	Lot. 49 No. of tubes 7 Hours req. 48 Spec. grav.	Vol. cc.	Lot. 50 No. of tubes 5 Hours req. 72—4 tubes 89—1 tube Spec. grav.	Vol. cc.
A	0.8385	125	0.8341	255	0.8320	170
B_1	0.8530	275	0.8505	395	0.8485	230
B_2	....	...	0.8520	95	0.8500	70
C_1	0.8568	320	0.8560	380	0.8565	300
C_2	0.8586	90	0.8572	230	0.8577	100
D_1	0.8610	325	0.8620	500	0.8609	480
D_2	0.8623	115	0.8625	290	0.8626	125
EF_1	0.8695	330	0.8705	500	0.8685	640
EF_2	0.8700	80	0.8705	580	0.8700	235
		1660		3225		2350

Observations on the Second Fractionation.

Specific Gravity.—The range of the specific gravities grows smaller as the oils to be fractionated become lighter, and less complex. Thus, in Lot 36, the range of specific gravity extends from 0.8272, the value for Fraction A, to 0.8356, the

[1] Owing to the weakness of the water pressure, the pressure in the tubes was only slightly diminished. The tubes were taken down before the reservoirs were completely exhausted. The distances to which the oil had risen were 35, 25, 30, 20, 10 cm. from the tops of the tubes.

[2] Owing to the weakness of the water pressure, the pressure in the tubes was diminished but slightly during this time. The tubes were taken down before the reservoirs were completely exhausted. The distances to which the oil had risen were 50, 35, 30, 60, 55 cm. from the tops of the tubes.

value for EF_2, the difference between them being 0.0084. In Lot 38, the mother oil, of specific gravity 0.8433, yielded fractions whose specific gravities ranged from 0.8331 to 0.8518, amounting to a difference of 0.0187. This fact appears to be general throughout the various lots, and points to the gradual formations of mixtures which will pass through the earth unaltered, just as the fractionation by distillation tends to yield substances with definite boiling points.

Color.—The color of the oils in this fractionation shaded from a very light yellow to greenish black.

Odor.—The odor of the crude petroleum vanished completely from the oils of this fractionation.

Volume of Oil Retained by the Earth.—The oil retained by the earth in this fractionation amounted to approximately 50 per cent., a smaller percentage, as is naturally to be expected, than in the fractionation of the crude petroleum.

The Third Fractionation.

The following oils obtained from the second fractionation were united for the third fractionation:

Lot 51.—Specific Gravity 0.8316.

Lot.	Fraction.	Specific gravity.	Volume, cc.	Lot.	Fraction.	Specific gravity.	Volume, cc.
36	A	0.8272	160	42	A	0.8325	210
39	A	0.8290	255	44	A	0.8330	85
37	A	0.8292	135	36	B_2	0.8331	58
40	A	0.8305	380	38	A	0.8331	180
36	B_1	0.8315	216	46	A	0.8332	210
41	A	0.8316	235	36	C_1	0.8334	350
50	A	0.8320	170	49	A	0.8341	255

2899

Lot 52.—Specific Gravity 0.8343.

Lot.	Fraction.	Specific gravity.	Volume, cc.	Lot.	Fraction.	Specific gravity.	Volume, cc.
36	D_1	0.8330	360	36	EF_1	0.8347	720
36	D_2	0.8339	320	36	EF_2	0.8356	320
47	A	0.8340	145	36	C_2	0.8355	85

1950

Lot 53.—Specific Gravity 0.8433.

Lot.	Fraction.	Specific gravity.	Volume, cc.	Lot.	Fraction.	Specific gravity.	Volume, cc.
45	A	0.8362	170	38	B_1	0.8447	175
48	A	0.8385	125	40	B_2	0.8453	155
37	B_1	0.8421	215	38	B_2	0.8455	210
39	B_1	0.8432	355	39	B_2	0.8458	50
40	B	0.8438	515				

1970

Lot 54.—Specific Gravity 0.8473.

Lot.	Fraction.	Specific gravity.	Volume, cc.	Lot.	Fraction.	Specific gravity.	Volume, cc.
39	B_2	0.8458	60	50	B_1	0.8485	230
41	B_1	0.8460	290	42	B_1	0.8487	265
37	C_1	0.8467	295	39	C_1	0.8492	455
41	B_2	0.8480	65	38	C_1	0.8490	305

1965

Lot 55.—Specific Gravity 0.8485.

Lot.	Fraction.	Specific gravity.	Volume, cc.	Lot.	Fraction.	Specific gravity.	Volume, cc.
37	D_1	0.8468	340	37	EF_2	0.8489	215
37	D_2	0.8485	152	38	D_1	0.8492	400
37	EF_1	0.8480	535	47	B_1	0.8500	275

1917

Lot 56.—Specific Gravity 0.8508.

Lot.	Fraction.	Specific gravity.	Volume, cc.	Lot.	Fraction.	Specific gravity.	Volume, cc.
50	B_2	0.8500	70	45	B_1	0.8510	210
49	B_1	0.8505	395	39	C_2	0.8513	180
44	B_1	0.8505	175	42	B_2	0.8515	54
46	B_2	0.8505	50	40	C_1	0.8518	600
38	C_2	0.8505	175				

1909

Lot 57.—Specific Gravity 0.8509.

Lot.	Fraction.	Specific gravity.	Volume, cc.	Lot.	Fraction.	Specific gravity.	Volume, cc.
38	D_1	0.8505	740	38	D_2	0.8509	295
39	EF_1	0.8508	710	38	EF_2	0.8518	355

2100

Lot 58.—Specific Gravity 0.8558.

Lot.	Fraction.	Specific gravity.	Volume, cc.	Lot.	Fraction.	Specific gravity.	Volume, cc.
49	B_2	0.8520	95	49	C_1	0.8560	380
45	B_2	0.8522	80	45	C_1	0.8562	265
41	C_1	0.8523	375	50	C_1	0.8565	300
48	B_1	0.8530	275	42	C_2	0.8567	56
40	C_2	0.8539	170	46	C_2	0.8567	95
42	C_1	0.8540	335	48	C_1	0.8568	320
41	C_2	0.8540	100	49	C_2	0.8572	230
47	C_1	0.8553	320	43	C_1	0.8575	200
46	C_1	0.8554	300				
							3896

Lot 59.—Specific Gravity 0.8563.

Lot.	Fraction.	Specific gravity.	Volume, cc.	Lot.	Fraction.	Specific gravity.	Volume, cc.
39	EF_1	0.8546	166	41	D_2	0.8571	110
40	D_1	0.8550	685	42	D_1	0.8572	420
41	D_1	0.8558	470	45	D_2	0.8580	100
39	EF_2	0.8560	350	42	D_2	0.8582	175
40	D_2	0.8560	330	48	C_2	0.8586	90
45	D_1	0.8567	425	47	D_1	0.8595	430
							4750

Lot 60.—Specific Gravity 0.8615.

Lot.	Fraction.	Specific gravity.	Volume, cc.	Lot.	Fraction.	Specific gravity.	Volume, cc.
46	D_1	0.8600	370	41	EF_1	0.8620	580
49	EF_1	0.8605	780	44	D_2	0.8620	120
43	D_1	0.8605	220	49	D_1	0.8620	500
44	D_1	0.8605	195	41	EF_2	0.8622	320
50	D_1	0.8609	480	48	D_2	0.8623	115
48	D_1	0.8610	325	49	D_2	0.8625	290
46	D_2	0.8613	120	50	D_2	0.8626	125
47	D_2	0.8618	70	42	E_1	0.8640	675
40	EF_2	0.8620	600				
							5880

Lot 61.—Specific Gravity 0.8680.

Lot.	Fraction.	Specific gravity.	Volume, cc.	Lot.	Fraction.	Specific gravity.	Volume, cc.
42	EF_2	0.8650	200	46	EF_2	0.8680	130
43	EF_1	0.8650	225	44	EF_2	0.8680	175
45	EF_1	0.8659	615	50	EF_1	0.8685	640
47	EF_1	0.8665	330	48	EF_1	0.8695	330
46	EF_1	0.8666	610	50	EF_2	0.8700	235
47	EF_2	0.8670	215	49	EF_1	0.8705	500
45	EF_2	0.8670	150	49	EF_2	0.8705	580
44	EF_1	0.8672	240				
							4975

The oils thus united were fractionated by fuller's earth again, with the results given in Table VI.

Table VI.—The Third Fractionation.

Lot.	51		52		53		54	
No. of tubes.	31		2		2		2	
Hours req.	60		60		48		48	
Frac.	Spec. grav.	Vol., cc.	Spec. grav.	Vol., cc.	Spec. grav.	Vol., cc.	Spec. grav.	Vol., cc.
A	0.8213	92	0.8219	65	0.8266	73	0.8303	66
B	0.8303	185	0.8333	143	0.8431	115	0.8488	115
C_1	0.8337	165	0.8375	190	0.8464	175	0.8518	175
C_2	0.8345	90	....	...	....	...	....	...
D_1	0.8353	210	0.8388	188	0.8468	145	0.8523	160
D_2	0.8356	170	0.8393	90	0.8474	115	0.8528	105
E_1	0.8366	385	0.8403	175	0.8473	202	0.8530	245
E_2	....	...	0.8411	92	0.8488	73	0.8548	60
F_1	0.8373	190	0.8431	88	0.8496	170	0.8548	145
		1487		1031		1068		1091

Lot.	55		56		57		58	
No. of tubes.	2		2		2		4	
Hours req.	48— 1 tube 72— 1 tube		96		96		72— 3 tubes 92— 1 tube	
Frac.	Spec. grav.	Vol., cc.	Spec. grav.	Vol., cc.	Spec. grav.	Vol., cc.	Spec. grav.	Vol., cc.
A	0.8283	58	0.8313	·75	0.8336	55	0.8318	170
B	0.8457	100	0.8488	135	0.8491	130	0.8531	260
C_1	0.8515	155	0.8546	.170	0.8528	180	0.8578	205
C_2	....	...	....	...	....	...	0.8592	105
D_1	0.8521	220	0.8553	150	0.8551	185	0.8588	205
D_2	0.8543	50	0.8560	92	0.8573	45	0.8593	340
E_1	0.8540	270	0.8553	145	0.8568	170	0.8603	325
E_2	....	...	0.8563	90	0.8588	70	0.8613	170
F	0.8566	180	0.8575	130	0.8611	170	0.8628	275
		1033		987		1005		2055

[1] The tin tubes used in these lots were 1.5 inches in diameter.
[2] The pressure in the tubes was diminished intermittently.

Lot.	59		60		61	
No. of tubes.	5		6		5	
Hours req.	72		72		5 days.[1]	
Frac.	Spec. grav.	Vol., cc.	Spec. grav.	Vol., cc.	Spec. grav.	Vol., cc.
A	0.8328	195	0.8343	195	0.8413	...
B	0.8508	340	0.8540	330	0.8601	...
C_1	0.8578	325	0.8601	290	0.8683	...
C_2	0.8588	112	0.8618	130	....	...
D_1	0.8608	490	0.8628	440	0.8709	...
D_2	0.8623	135	0.8638	85	....	...
E_1	0.8628	475	0.8664	425	0.8688	...
E_2	0.8633	155	0.8683	140	....	...
F	0.8673	330	0.8703	310	0.8691	...

Observations on the Third Fractionation.

Specific Gravity.—The decrease in the range of specific gravity as the oils supplied become lighter is observed in this fractionation as in the preceding ones.

Color.—The lightest oils were almost colorless; the heavier oils were dark brown to green.

Odor.—Most of the oils possessed an agreeable odor.

Prolonged Diffusion.—In Lot 61, the time required for the oils to reach the tops of the tubes was five days. No fractionation, as is evident from an examination of the specific gravities, occurred in the lower parts of the tubes. The heavier oils of fractions D, E, and F were exceedingly viscous.

Volume of Oil Retained by the Earth.—The volume of oil retained by the earth in this fractionation amounted to approximately 45 per cent. The increase in the yield of oil indicates, therefore, a process of purification, in which, as will be shown later, such compounds as the unsaturated hydrocarbons are removed.

The Fourth Fractionation.

The following fractions obtained from the third fractionation were united for the fourth fractionation:

[1] See below, this page.

Lot 62.—Specific Gravity 0.8298.

Lot.	Fraction.	Specific gravity.	Volume, cc.	Lot.	Fraction.	Specific gravity.	Volume, cc.
51	A	0.8213	92	51	B	0.8303	185
52	A	0.8219	65	56	A	0.8313	75
53	A	0.8266	73	58	A	0.8318	170
55	A	0.8283	66	59	A	0.8328	195
54	A	0.8303	58				
							979

Lot 63.—Specific Gravity 0.8343.

Lot.	Fraction.	Specific gravity.	Volume, cc.	Lot.	Fraction.	Specific gravity.	Volume, cc.
52	B	0.8333	143	51	C_2	0.8345	90
57	A	0.8336	55	31	D_1	0.8353	210
51	C_1	0.8337	185	51	D_2	0.8356	170
60	A	0.8343	195				
							1048

Lot 64.—Specific Gravity 0.8368.

Lot.	Fraction.	Specific gravity.	Volume, cc.	Lot.	Fraction.	Specific gravity.	Volume, cc.
51	E_1	0.8366	388	52	C_1	0.8375	190
51	F	0.8372	190	52	D_1	0.8388	188
							956

Lot 65.—Specific Gravity 0.8430.

Lot.	Fraction.	Specific gravity.	Volume, cc.	Lot.	Fraction.	Specific gravity.	Volume, cc.
52	D_2	0.8393	90	52	F	0.8431	88
52	E_1	0.8403	175	55	B_1	0.8457	100
52	E_2	0.8411	92	53	C_1	0.8464	175
53	B_1	0.8431	115	53	D_1	0.8468	145
							980

Lot 66.—Specific Gravity 0.8483.

Lot.	Fraction.	Specific gravity.	Volume, cc.	Lot.	Fraction.	Specific gravity.	Volume, cc.
53	E_1	0.8473	202	56	B_1	0.8488	135
53	D_2	0.8474	115	53	E_2	0.8488	73
54	B_1	0.8488	115	59	B_1	0.8508	330
							970

Lot 67.—Specific Gravity 0.8513.

Lot.	Fraction.	Specific gravity.	Volume, cc.	Lot.	Fraction.	Specific gravity.	Volume, cc.
57	B_1	0.8491	130	54	C_1	0.8518	175
59	B_1	0.8508	10	55	D_1	0.8521	220
55	C_1	0.8515	155	58	B_1	0.8531	260
							950

Lot 68.—Specific Gravity 0.8533.

Lot.	Fraction.	Specific gravity.	Volume, cc.	Lot.	Fraction.	Specific gravity.	Volume, cc.
54	D_1	0.8523	180	54	E_1	0.8530	245
54	D_2	0.8528	105	60	B	0.8540	330
57	C_1	0.8528	180				
							1040

Lot 69.—Specific Gravity 0.8556.

Lot.	Fraction.	Specific gravity.	Volume, cc.	Lot.	Fraction.	Specific gravity.	Volume, cc.
55	E_1	0.8540	270	56	E_1	0.8553	145
55	D_2	0.8543	50	56	D_2	0.8560	92
56	C_1	0.8546	170	56	E_2	0.8563	90
54	E_2	0.8548	60	55	F	0.8566	180
54	F	0.8548	145	57	E_1	0.8568	170
57	D_1	0.8551	185	57	D_2	0.8573	45
56	D_1	0.8553	150	56	F	0.8575	130
							1882

Lot 70.—Specific Gravity 0.8596.

Lot.	Fraction.	Specific gravity.	Volume, cc.	Lot.	Fraction.	Specific gravity.	Volume, cc.
58	C_1	0.8578	205	60	C_1	0.8601	290
59	C_1	0.8578	325	58	E_1	0.8603	325
58	D_1	0.8588	205	59	D_1	0.8608	490
59	C_2	0.8588	112	57	F	0.8611	170
57	E_2	0.8588	70	58	E_2	0.8613	170
58	C_2	0.8592	105	60	C_2	0.8618	130
58	D_2	0.8593	340				
							2937

Lot 71.—Specific Gravity 0.8638.

Lot.	Fraction.	Specific gravity.	Volume cc.	Lot.	Fraction.	Specific gravity.	Volume, cc.
59	D_2	0.8623	135	60	D_2	0.8638	85
60	D_1	0.8628	440	59	E_2	0.8633	155
59	E_1	0.8628	475	60	E_1	0.8664	425
58	F	0.8628	375				
							1990

Table VII.—The Fourth Fractionation.

	Lot. 62		Lot. 63		Lot. 64		Lot. 65	
No. of tubes.	1		1		1		1	
Hours req.	72		72		90		48	
Frac.	Spec. grav.	Vol., cc.	Spec. grav.	Vol., cc.	Spec. grav.	Vol., cc.	Spec. grav.	Vol., cc.
A	0.8243	32	0.8273	45	0.8297	41	0.8308	42
B	0.8298	71	0.8357	75	0.8378	57	0.8428	70
C	0.8323	90	0.8378	95	0.8401	81	0.8463	92
D	0.8330	115	0.8383	130	0.8408	115	0.8473	130
E	0.8333	130	0.8388	98	0.8413	135	0.8471	130
F	0.8341	75	0.8393	95	0.8418	70	0.8485	80
		513		538		499		544

Observations on the Fourth Fractionation.

Specific Gravity.—As in the preceding fractionations, the decrease in the range of specific gravity as the mother oils become lighter is again observed in this fractionation. It is evident, moreover, that there is a constant forward accumulation towards definite and constant mixtures. The lighter oils of one lot are found to possess specific gravities closely approaching those of the heavier oils of the preceding lot.

Color.—The oils of Fraction A were almost colorless; the color of the heavier oils ranges from green to light brown.

Odor.—All the oils of this fractionation possessed agreeable odors.

Volume of Oil Retained.—The volume of oil retained by the earth amounted to approximately 40 per cent.

Deposition of Paraffin.—In Fractions A and B of several of the lots, a fine, crystalline deposit separated out, and collected upon the bottom of the bottles containing the oils. When the oils were warmed, this deposit dissolved completely, showing it to be paraffin.

Chemical Examination of the Fractionated Oils. Unsaturated Hydrocarbons.

Action of Concentrated Sulphuric Acid.—The percentage of volume of oil absorbed by concentrated sulphuric acid (specific gravity 1.84) was determined according to the following procedure: Ten cc. of the oil to be examined were measured into a glass-stoppered bottle, and thirty cc. of concentrated sulphuric acid were added. The mixture was thoroughly shaken in a machine for thirty minutes and then poured into a burette. After allowing sufficient time for any oil that might be mechanically absorbed in the acid to rise to the top, the volume of unabsorbed oil was read directly over the acid. Owing to the formation of heavy emulsions, no attempt was made to neutralize and wash the oil.

The results of the analyses are given in the table below:

Lot.	Fraction.	Per cent. by volume absorbed.	Lot.	Fraction.	Per cent. by volume absorbed.
51	A	2.3	51	D_1	11.5
51	B	6.1	51	D_2	12.0
51	C_1	9.1	51	E	12.5
51	C_2	10.2	51	F	14.5

Action of Bromine.—The method employed for determining the amount of bromine absorbed by the oils was as follows: Between 0.5 and 0.9 of a gram of the oil to be examined was dissolved in 10 to 15 cc. of carbon tetrachloride. Five cc. of a standard solution of bromine in carbon tetrachloride were then introduced, and the solution allowed to remain, with occasional shaking, in a dark place for 30 minutes. Ten cc. of a 10 per cent. solution of potassium iodide were then added, and the amount of iodine liberated determined immediately by titrating with a standard solution of sodium thiosulphate. A few drops of a starch solution were introduced to mark accurately the end of the titration. The amounts of bromine absorbed by addition and substitution were not estimated separately.

The amount of bromine absorbed, calculated upon the basis of 100 grams of oil, are given in the following table:

Table VIII.—First Fractionation.

Lot.	Fraction.	Per cent. of bromine absorbed.	Lot.	Fraction.	Per cent. of bromine absorbed.
32	A	5.02	32	D	7.87
32	B	6.96	32	E	8.00
32	C	7.40	Crude oil		7.64

Second Fractionation.

Lot.	Fraction.	Per cent. of bromine absorbed.	Lot.	Fraction.	Per cent. of bromine absorbed.
36	A	4.74	36	D_1	6.81
36	B_1	5.40	36	D_2	6.28
36	B_2	5.66	36	EF_1	6.49
36	C_1	5.56	36	EF_2	7.18
36	C_2	6.18			

Third Fractionation.

Lot.	Fraction.	Per cent. of bromine absorbed.	Lot.	Fraction.	Per cent. of bromine absorbed.
51	A	3.27	51	D	4.92
51	B	4.36	51	E	4.71
51	C	4.47	51	F	5.36

Fourth Fractionation.

Lot.	Fraction.	Per cent. of bromine absorbed.	Lot.	Fraction.	Per cent. of bromine absorbed.
62	A	2.86	62	E	3.73

These results demonstrate conclusively that the unsaturated hydrocarbons tend to collect in the lower sections of a layer of fuller's earth through which the oil is allowed to diffuse. These figures confirm the results obtained by Gilpin and Cram in their work on the Pennsylvania petroleum. In their investigation, distillation by heat was employed in order to obtain fractions that could be readily studied. In this work the relative amounts of the unsaturated hydrocarbons in the various oils were determined directly as they came from the earth.

The percentages by volume of oil absorbed by concentrated sulphuric acid represent only approximately the percentages of unsaturated hydrocarbons, since, as was shown previously, any benzene which may have been present in the oils was also removed by the concentrated acid. This fact rendered impossible a quantitative separation of the aromatic from the unsaturated hydrocarbons. Since no other methods, besides nitration and sulphonation, neither of which could be here employed, were available, no results as to the relative amounts of the aromatic hydrocarbons in the various fractions could be obtained.

It is evident from the results of the bromine determinations that, as the fractionation proceeds, the amounts of unsaturated hydrocarbons become smaller and smaller. A comparison of the amounts of bromine absorbed by Fraction A of the first, second, third and fourth fractionations is given below for the purpose of bringing out this point more clearly:

Per Cent. of Bromine Absorbed by Fraction A.

First fractionation.	Second fractionation.	Third fractionation.	Fourth fractionation.
5.02	4.74	3.27	2.86

Sulphur Compounds.—The sulphur in the various oils was determined by the usual method of combustion. For these determinations, the oils obtained from one tube of Lot 6 were employed. The results are given in the following table:

Lot 6 Fraction.	Specific gravity.	Per cent. sulphur.	Lot 6. Fraction.	Specific gravity.	Per cent. sulphur.
A	0.8195	0.04	D	0.8510	0.09
B	0.8362	0.05	E	0.8600	0.16
C	0.8440	Lost			

The percentage of sulphur in the Fractions A, C and E of Lot 51 was also determined. The results were as follows:

Table X.—Lot 51.

Fraction.	Per cent. of sulphur.	Fraction.	Per cent. of sulphur.
A	0.003	E	0.006
C	0.040		

These results show that the sulphur tends to collect in the oils in lower sections of the tube. As the fractionation proceeds, the proportion of sulphur becomes smaller. The figures below indicate that as the oil is subjected to repeated fractionations, the sulphur is gradually removed:

Per Cent. of Sulphur.

	First fractionation.	Second fractionation.	Third fractionation.
Fraction A	0.04	....	0.003
Fraction E	0.16	....	0.006
Fraction C	....	0.08	0.040

Selective Action of Fuller's Earth.

When the earth, from which as much oil as possible has been extracted by prolonged treatment of water, is dried, and then digested with ether, oils of surprisingly high specific gravity and viscosity are obtained.

In the experiments undertaken to study the selective action of fuller's earth, the procedure was as follows: The earth under examination was thoroughly treated with water until no more oil appeared. This muddy earth of the consistency of thin liquid paste was spread upon porous plates and allowed to dry at room temperature. Several weeks usually elapsed before the earth became completely dry. It was then finely pulverized, and after being thoroughly soaked and shaken with ether the mixture was allowed to remain undisturbed for 24 hours or more. The mixture was then filtered and the dissolved oil recovered by distilling off the ether from the filtrate. The residual earth was then digested with ether for some time by means of an electric stove that completely surrounded the flask. The oil thus extracted was added to the oil first obtained. In several cases the residual earth was treated further with ether in the Soxhlet extractor.

The results of these extractions are given in the following table:

Table XI.

Lot.	Fraction.	Specific gravity at 50°.	Lot.	Fraction.	Specific gravity at 50°.
7	A	0.8470	25	A_2	0.8391
8	A	0.8502	25	B	0.8489
18	A_1	0.8419			Specific gravity at 20°.
18	A_2	0.8400	51	A	0.8368
19	A_1	0.8495	51	B	0.8473
19	A_2	0.8495	51	C	0.8491
19	A_3	0.8600	71	D	0.8568
25	A_1	0.8363	51	E	0.8518
25	A_2	0.8381	51	F	0.8553

The specific gravity of none of the oils of the first and second fractionation, extracted with ether, except those of Lot 19, could be determined at 20° C. All were extremely viscous; those of Lot 25 were so viscous at this temperature that they would not flow when the bottles containing them were inclined. The color of the oils ranges from brown to black. The ethereal solutions, however, of many of the oils were very light in color.

It is interesting to compare the specific gravities of the oils extracted with ether with those of the corresponding oils extracted with water. For this purpose, the oils extracted by water and by ether from the earth of Lot 51 are chosen. In the table below, the specific gravities of these oils at the same temperature, i. e., 20°, are given:

Table XII.—Lot 51. Specific Gravity at 20°.

Frac.	Ether.	Water.	Frac.	Ether.	Water.
A	0.8368	0.8213	D	0.8568	0.8353
B	0.8473	0.8303	E	0.8518	0.8366
C	0.8491	0.8337	F	0.8553	0.8373

As the figures above indicate, the specific gravities of oils extracted with ether are much higher than those of the corresponding oils extracted with water. The presence of such heavy and viscous oils in the upper sections of the tube can be explained only by assuming that they were carried to these heights in solution with the lighter oils, and were then removed by the earth. Since such viscous oils are totally unable to diffuse by capillarity to any appreciable extent, it is not probable that their transportation to the upper parts of the tube was effected by capillary diffusion.

Chemical Examination of the Oils Extracted by Ether—Unsaturated Hydrocarbons.

Action of Concentrated Sulphuric Acid.—The percentage by volume of oil absorbed by concentrated sulphuric acid (specific gravity 1.84) was determined according to the following procedure: Ten cc. of the oil to be examined were measured into a glass-stoppered bottle, and thirty cc. of concentrated sulphuric acid were added. The mixture was thoroughly shaken in a machine for 30 minutes and then poured into a burette.

After allowing sufficient time for any oil that might be mechanically absorbed in the acid to rise to the top, the volume of
unabsorbable oil was read directly over the acid. Owing to
the formation of heavy emulsions, no attempt was made to
neutralize and wash the oil.

The oils selected for examination were those extracted by
ether from the earth of Lots 36 and 51.

The results of the analyses are given in the following table:

Table XIII.

Lot.	Fraction.	Oils extracted by ether. Per cent. by volume absorbed.	Oils extracted by water. Per cent. by volume absorbed.	Lot.	Fraction.	Oils extracted by ether. Per cent. by volume absorbed.	Oils extracted by water. Per cent. by volume absorbed.
36	A	24.0	3.0	51	C	17.0	9.1
36	B	37.0	10.4	51	D	16.4	11.5
51	A	7.0	2.3	51	E	16.5	12.5
51	B	11.5	6.1	51	F	18.0	14.5

Action of Bromine.—The method employed for determining
the amount of bromine absorbed by the oils has already been
described (p. 50).

The amounts of bromine absorbed, calculated upon the
basis of 100 grams of oil, are given in the table below. The
values for the corresponding oils extracted with water are
also given for comparison.

Table XIV.

Lot.	Fraction.	Oils extracted by ether. Per cent. of bromine absorbed.	Oils extracted by water. Per cent. of bromine absorbed.	Lot.	Fraction.	Oils extracted by ether. Per cent. of bromine absorbed.	Oils extracted by water. Per cent. of bromine absorbed.
32	A	5.30	5.02	51	B	4.45	4.36
32	B	7.39	6.96	51	C	6.27	5.03
36	A	5.72	4.74	51	D	6.09	4.92
36	B	6.10	5.40	51	E	5.98	4.71
36	C	6.72	5.56	51	F	5.20	5.36
51	A	3.27	3.27				

As these results clearly demonstrate, fuller's earth retains
the unsaturated hydrocarbons, thus exercising a selective
action.

Sulphur Compounds.—The sulphur in the oils obtained by extraction with ether was determined by the usual method of combustion. The results are given in the table below:

Table XV.

Lot. 51.	Oils extracted by ether. Per cent. of sulphur.	Oils extracted by water. Per cent. of sulphur.	Lot 51.	Oils extracted by ether. Per cent. of sulphur.	Oils extracted by water. Per cent. of sulphur.
A	0.004	0.003	D	0.060	...
B	0.011	...	E	0.080	...
C	0.050	0.040	F	0.080	0.006

The selective action of the earth, in regard to the sulphur compounds, is indicated by these results. This fact was also pointed out by Richardson and Wallace. It is very probable that the earth also retains largely the nitrogen compounds in the oil, and may also remove to a greater or less extent the benzene hydrocarbons.

These results seem to furnish evidence in favor of the view that the Pennsylvania oil diffused, at some time in the history, through porous media, which exercised a selective action upon it, removing a large part of the unsaturated and sulphur compounds, and probable the benzene and nitrogen compound.

SUMMARY.

1. When a solution of benzene and a paraffin oil is allowed to diffuse upward through a tube packed with fuller's earth, the benzene tends to collect in the lower sections and the paraffin oil in the upper sections of the tube.

2. When crude petroleum diffuses upward through a tube packed with fuller's earth, a fractionation of the oil occurs. The oil that is displaced by water from the earth from the top of the tube possesses a lower specific gravity than the oil obtained from the earth at the bottom of the tube.

3. As the fractionation proceeds, the range of specific gravity covered in succeeding fractionations becomes smaller, indicating a movement towards the production of mixtures which will finally pass through the earth, unaltered.

4. In the fractionation of petroleum by capillary diffusion through fuller's earth, the amounts of unsaturated hydrocarbons and sulphur compounds in the resulting fractions increase gradually from the lightest oils at the top to the heavier oils at the bottom of the tube.

5. Fuller's earth tends to retain the unsaturated hydrocarbons and sulphur compounds in petroleum, thus exercising a selective action upon the oil.

BIOGRAPHICAL.

The author of this dissertation, Oscar Ellis Bransky, was born in Baltimore, January 29, 1886. He received his early education in the public schools of Baltimore. In 1904, he graduated from the Baltimore City College, and in October of the same year he entered the Johns Hopkins University. He received the degree of Bachelor of Arts in 1907. He then entered the post-graduate department of the university, pursuing Chemisty as his major, and Physical Chemistry and Geology as his minor subjects. During 1908–1909 he acted as laboratory assistant to Professor Renouf.